I0015372

Using FREE Scribus Software to Create Professional Presentations

Book Covers, Magazine Covers, Graphic Designs, Posters, Newsletters, Renderings, and More (Full Color Edition)

Alice Chen & Gang Chen

ArchiteG®, Inc.
Irvine, California

Using FREE Scribus Software to Create Professional Presentations: Book Covers, Magazine Covers, Graphic Designs, Posters, Newsletters, Renderings, and More (Full Color Edition)

Copyright © 2010 Gang Chen
V1.0
Cover Photo © 2010 Gang Chen

Copy Editor: Penny L Kortje

All Rights Reserved.
No part of this book may be transmitted or reproduced by any means or in any form, including electronic, graphic, or mechanical, without the express written consent of the publisher or authors, except for cases of brief quotation in a review.

ArchiteG®, Inc.
http://www.ArchiteG.com

ISBN: 978-0-9843741-5-1

PRINTED IN THE UNITED STATES OF AMERICA

Dedication

To my grandpa Zhuixian and grandma Yugen,
my mom, Xiaojie, and my sisters,
Angela, Amy, and Athena.

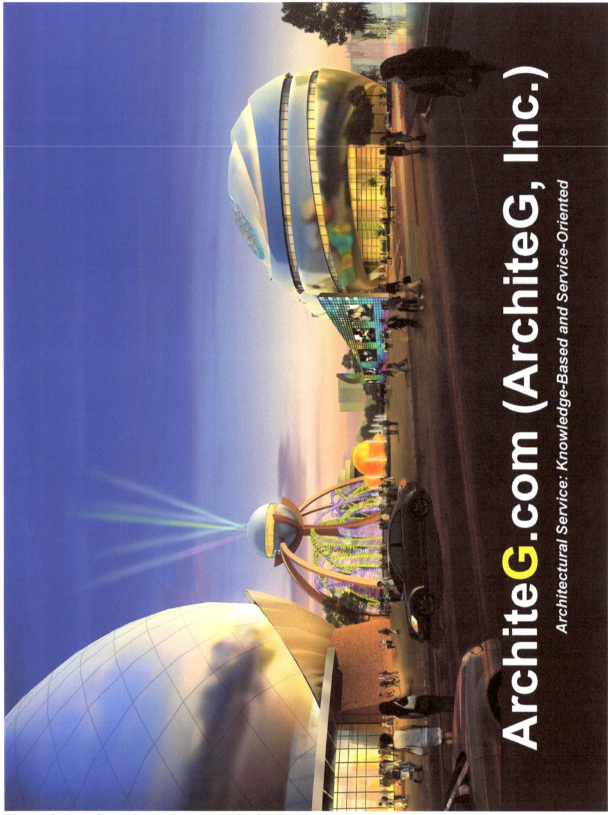

Figure 0.1   Sample poster created with Scribus

Disclaimer

Using FREE Scribus Software to Create Professional Presentations provides general information about Scribus. The book is sold with the understanding that neither the publisher nor the authors are providing legal, accounting, or other professional services. If legal, accounting, or other professional services are required, seek the assistance of a competent professional firm.

The purpose of this publication is not to reprint the content of all other available texts on the subject. You are urged to read other materials, and tailor them to fit your needs.

Great effort has been taken to make this resource as complete and accurate as possible. However, nobody is perfect and there may be typographical errors or other mistakes present. You should use this book as a general guide and not as the ultimate source on this subject. If you find any potential errors, please send an e-mail to:
info@ArchiteG.com

Using FREE Scribus Software to Create Professional Presentations is intended to provide general, entertaining, informative, educational, and enlightening content. Neither the publisher nor the authors shall be liable to anyone or any entity for any loss or damages, or alleged loss or damages, caused directly or indirectly by the content of this book.

If you do not wish to be bound by the above, you may return this book to the publisher for a full refund.

Legal Notice

Figure 0.2   Sample book cover created with Scribus

How to Use This Book

We suggest you read *Using FREE Scribus Software to Create Professional Presentations* at least three times:

Read once all the way through to highlight unfamiliar or important information.

Read twice focusing on the highlighted information to memorize. You can repeat this process as many times as you want until you master the content of the book.

The Table of Contents is very detailed so you can locate information quickly. If you are on a tight schedule you can forgo reading the book linearly and jump around to the sections you need. To make your tasks easier, we included concise command descriptions and straightforward color screen shots. After you read the book once, you should know enough to simply read the **bolded command lines** and look through the **screen shots**, to perform tasks.

All commands are described in an **abbreviated manner**. For example, **View > Fit to Width** means go to the pull down menu on the top of your computer screen, click **View,** and then click **Fit to Width.** This is typical for ALL pull down menu commands throughout the book.

This book is designed as a **turnkey operation manual**. Since book cover design is one of the most difficult forms of professional presentations, we focused on this type of design to document the procedure of creating professional presentations step by step. Anyone with a seventh grade education should be able to consistently produce book covers or apply the information to create other types of professional presentations.

LEED GA
MOCK EXAMS

QUESTIONS, ANSWERS, AND EXPLANATIONS

A MUST-HAVE FOR THE LEED GREEN ASSOCIATE EXAM, GREEN BUILDING LEED CERTIFICATION, AND SUSTAINABILITY

LEED GA MOCK EXAMS

Gang Chen   ArchiteG.com

GANG CHEN

Construction - General

Pass the LEED Green Associate Exam, Get Your Building LEED Certified, Fight Global Warming, and Save Money!

LEED (Leadership in Energy and Environmental Design) is one of the most important trends in development and is revolutionizing the construction industry. It has gained tremendous momentum and has a profound impact on our environment.

From this book, you will be able to:

1. Identify your weakness through practice questions
2. Learn to work well under the pressure of timed tests
3. Check your responses against the solutions
4. Understand the solutions for the difficult questions through the explanations
5. Fully understand the scope, difficulty level, and format of the LEED Green Associate Exam
6. Learn how to pass the LEED Green Associate Exam.

There is NO official GBCI book on the LEED Green Associate Exam. *LEED GA Mock Exams* fills in the blanks and demystifies LEED. The book includes 200 questions and simulates the real exam in every aspect, including scope, difficulty level, format, and number of questions in each LEED category. It includes questions, answers, and explanations. This book is small and easy to carry around. You can read it whenever you have a few extra minutes. It is an indispensable resource for ordinary people, developers, brokers, contractors, administrators, architects, landscape architects, engineers, interns, drafters, designers, and other design professionals.

Gang Chen is a LEED AP BD+C and a licensed architect in California. He is also the internationally acclaimed author of other fascinating books, including: *Architectural Practice Simplified, Planting Design Illustrated,* and LEED Exam Guide series, which includes one guide book for each of the LEED exams.

ISBN 978-0-9843741-2-0

90000

9 780984 374120

ArchiteG.com

Figure 0.3   Sample book cover created with Scribus

Table of Contents

Chapter One   Preparation

Chapter Two   Setting up the Sheet and Preparing the Background

Chapter Three   Adding the Barcode, Images, Titles, and Text

Appendixes

Back Page Promotion

Index

Chapter One

Preparation

A. Downloading the FREE software

Before you start, make sure to download all of the software required. Below is a list of software required to create a book cover. Each item is followed by a link to where you can download the software for FREE.

1. **Adobe Reader** for viewing PDF files:
http://get.adobe.com/reader/

2. **Ghostscript**:
http://sourceforge.net/projects/ghostscript/

3. **OpenOffice.org**:
http://download.openoffice.org/other.html##en-US

OpenOffice.org is both the name of the software and the name of the organization that created it. OpenOffice.org is a great FREE substitute for **Microsoft Office Suite**. If you already have Microsoft Office Suite, you do NOT need to download.

4. **GIMP** or **PhotoScape** for editing or modifying images:
GIMP:
http://www.gimp.org/windows/

PhotoScape:
http://www.photoscape.org

We prefer PhotoScape because it is so easy to resize, crop, and trim JPEG images. You can even crop to round shapes or other free forms.

5. **Scribus:**
http://www.scribus.net/downloads

Make sure you download the latest **development** release (currently 1.3.8) instead of the latest **stable** release of Scribus. The stable release does NOT allow you to export and generate **PDF files**; the development release does. To print your final book cover, you NEED a PDF file. Therefore, the development release of Scribus is your ONLY choice.

After downloading and installing all of the software mentioned above, you may need to restart your computer for the software to be fully functional. Once you begin your book cover design, do NOT upgrade to any newer versions of the software until you finish. A newer version may create unintended consequences for your design, and require you to start the whole process again.

B. Getting the Cover Template

All major book printers have a FREE book cover template generator. You simply go to their website, fill in some basic information about your book, and can have a nice book template e-mailed to you.

Tips:
*Before you create your book's interior typesetting or book cover design, make sure to check your book printer's website for available book sizes. Your book interior layout and cover design must **match** one of the **available book sizes**. If you have been hired by a client to create a cover, consult with him about which book printer he uses and his desired size.*

Throughout this book, we will use my book *Building Construction* as an example.

Sample:
The following is the information required by our book printer's cover generator and the underlined information that we provided.

✓ ?**13-digit ISBN (with dashes):** 978-0-9843741-4-4 Click here to convert your 10-digit ISBN to a 13-digit ISBN

?**Publisher Reference Number:**_____(We suggest you leave this **blank.**)

Content Type: ◉ **B&W** ◯ **Color** (This is referring to your book's **interior**, NOT the book cover.)

Paper Type: ◯ **Creme** ◉ **White** (Again, this is referring to your book's **interior**, NOT the book cover.)

Note:
*We **always** use white paper since the print quality is MUCH **higher** than with Crème paper. This is based on our past experience and other people's suggestions.*

✓**Book Type:** 8.5 x11 in or 280 x215 mm Perfect Bound on white

✓ **Page Count:** 194 **(Multiple of 2, between 18 and 1200)**

Note:
The actual page count for the book is 193. We rounded up to 194, because a single sheet of paper is double sided and actually counts as two pages.

File Type to Return: PDF (You can choose PDF, EPS, InDesign CS3 and newer, and QuarkXpress 7 and newer. We suggest you request both PDF and EPS file formats.)

✓ **Email Address:** info@ArchiteG.com

✓ **Retype Email Address:** info@ArchiteG.com

Optional Information:

Price (including decimal): _____ (We suggest you **always** leave this **blank** because, if you fill in the blank and then change the price later, you will have to pay an **additional fee.**)

Currency: US Dollars

Price in Barcode: NO

After you complete the book printer's cover generator form and click **Submit**, a PDF book cover template (similar to figure 1.1) will be e-mailed to you. To request an EPS file, you can repeat the same process except change **File Type to Return** from PDF to EPS.

Make sure you **get both a PDF and EPS file**. The PDF file will provide you with the basic information needed to set up the book cover, while the EPS file can be **imported and used to extract information** such as a barcode, or **overlaid** with your book cover to double check that your text and images are within the limits of the template.

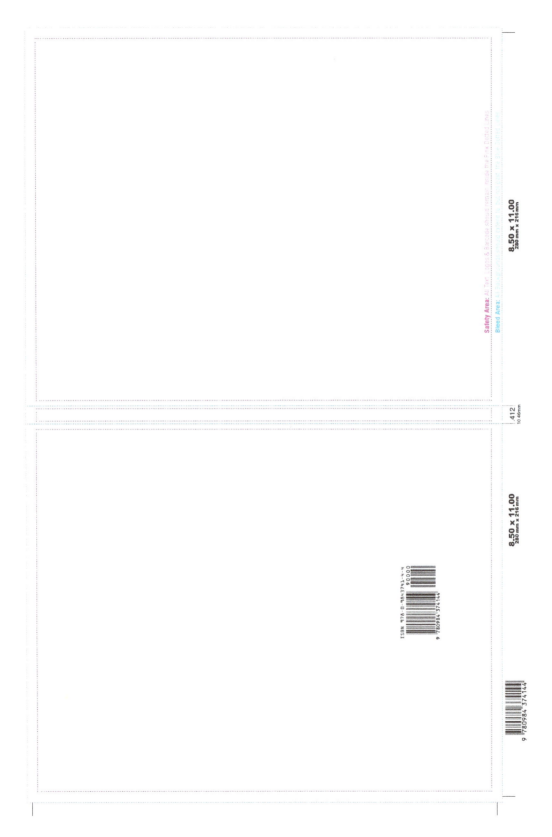

Figure 1.1   Part of the cover template created with a book printer's cover generator

A helpful way to think of a book cover is as a single sheet of paper composed of three basic elements: front cover, back cover, and the spine between. Before you create your book cover, you should become familiar with two important concepts that affect the size of layout you submit to the printer and the placement of text and images.

1. **Bleed area**

   When a book cover goes into the production phase there is an area along all four sides of the paper, which is eventually cut away to form the finished product. This is the **bleed area**. In other words, your **final** book cover is NOT the same size as the book cover **design**. The paper size for your design is the final book cover dimensions **PLUS the bleed area** which is normally **1/8"** or **0.125"** on ALL sides for most book printers.

   Note:
   *Check **the bleed area dimension** for **YOUR** specific book printer. It may not be 1/8". For example, some book printers use 1/8" bleed area for books with black and white interior, but use 1/4" bleed area for books with color interior.*

   If you do NOT include the bleed area in the PDF file that you submit to your book printer, the file will be rejected. You'll have to **revise** PLUS pay an **additional fee for resubmittal**.

   For our example (figure 1.1), the dimensions of the final book are as follows:

   Final book cover **width** = 8.5" (back cover width) + 0.412" (spine width) + 8.5" (front cover width) = **17.412"**

   Final book cover **height** = 11"

   Note:
   See the section on calculating the book cover design width and height in chapter two (page 30).

2. **Safety Area and Minimum Margin**

   No title, text, logo, barcode, or image that you wish to see in full should bleed to the edge of your book. They should ALL stay within the safety area. For most book printers, the **safety area** is the **final** book cover size **MINUS a minimum of 1/8" margin** for ALL four edges of the **final** front and back covers, **MINUS a minimum of 1/8" margin** for the top and bottom edges of the book spine, and **MINUS a minimum of 1/16" margin** for the left and right edges of the book spine.

   Note:
   *Check **the dimensions of minimum margin** for **YOUR** specific book printer. It may not be 1/8" or 1/16".*

Chapter Two

Setting up the Sheet and Preparing the Background

A. Setting up the basic dimensions of the book cover

Double-click the Scribus icon on your desktop, and a screen resembling figure 2.1 will pop up. Still using *Building Construction* as an example, we set the following underlined values detailed below and also depicted in figure 2.1.

On the right-hand side of the pop-up window complete as follows:
Default Unit: Inches (in)

Size: Custom

Orientation: Landscape

Width: 17.412 in

Height: 11 in

Number of Pages: 1

Note:
See the section on bleed area in chapter one (page 15) for information on how we came up with the width and height dimensions.

On the left-hand side of the pop-up window complete as follows:
Under Document Layout select Single Page.

Under **Margin Guides**
Left: 0.25 in
Right: 0.25 in
Top: 0.25 in
Bottom: 0.25 in

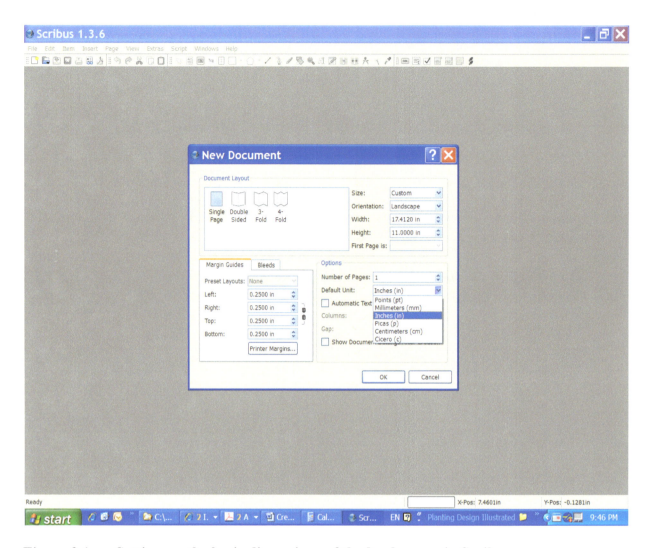

Figure 2.1   Setting up the basic dimensions of the book cover in Scribus

Click OK to close the pop-up window.

View > Fit to Width
File > Save As
Name the file (BC_Cover.sla).

*Important: ALWAYS save your file **every 15 minutes** to preserve your work.*

B. Adding the bleed area
    File > Document Setup
    Select **Bleeds** and fill in the following dimensions:
    Left: 0.125 in
    Right: 0.125 in
    Top: 0.125 in
    Bottom: 0.125 in

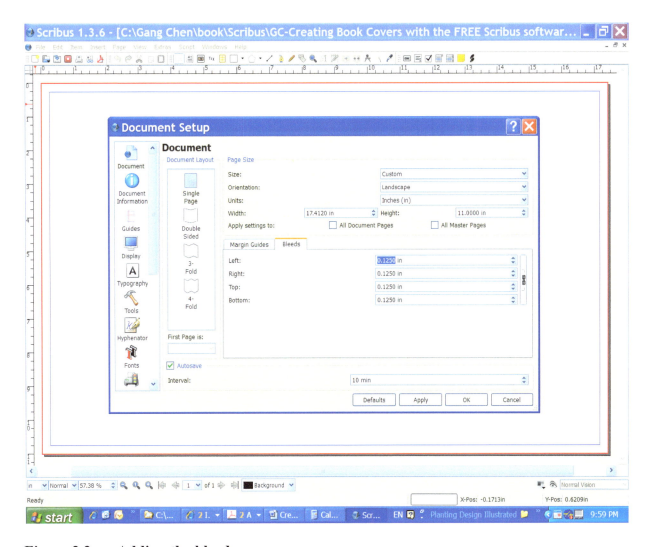

Figure 2.2   Adding the bleed area

C. Adding guides

To show the actual edges of the book spine add two vertical guides.

1. Adding the first guide

File > Manage Guides

A pop-up window appears. Select the **Single** tab and then the **Add** button on the right, under Verticals (in).

Fill in <u>8.5</u> under **Verticals (in)**.

2. Adding the second guide

Repeat step one, and fill in <u>8.912</u> under **Verticals (in)**.

Note:

The equation used to find this number is 8.5" (final back cover width) + 0.412" (spine width) = 8.912".

Your screen should look like figure 2.3.

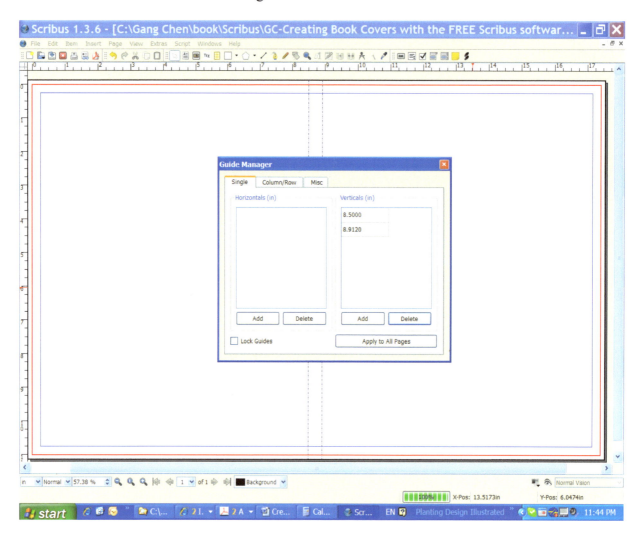

Figure 2.3   Adding guides

The **red** rectangle shows the **final** book size, the white edge outside the **red** rectangle is the 1/8" wide **bleed area**, and the two vertical blue dashed-dot lines in the middle show the EXACT locations of the book spine edges. The solid blue rectangle shows the margin guides depicted in figure 2.1.

You can add more vertical guides (like the centerlines of the front and back covers, or spine), or horizontal guides as necessary.

Click the "x" sign to close the pop-up window.

D. Setting up Layers

The Layers tool is very useful to organize information in Scribus and other graphic software. You can imagine layers as a stack of transparent paper. Each piece of paper is a layer that can be assigned particular information such as text, images, logos, or barcodes. The information on any layer will cover or superimpose the information on any layer below it. However, you can always turn a layer on and off or change the order of the layers. If a layer is turned off, the information assigned to that layer will NOT show up. In order to modify or add information, you need to go to the **particular** layer where you wish to make changes.

1. **Adding a new layer (figure 2.4)**
   Windows > Layers
   A pop-up window appears. **Click the "+" sign** and you'll get a new layer called "New Layer 1". Double-click "New Layer 1" to highlight, and then change the name to "Images".

2. **Repeat step one to create a layer with each of the following names:**
   Text
   Barcode
   Template
   Background Color

   You can create more layers later if you desire.

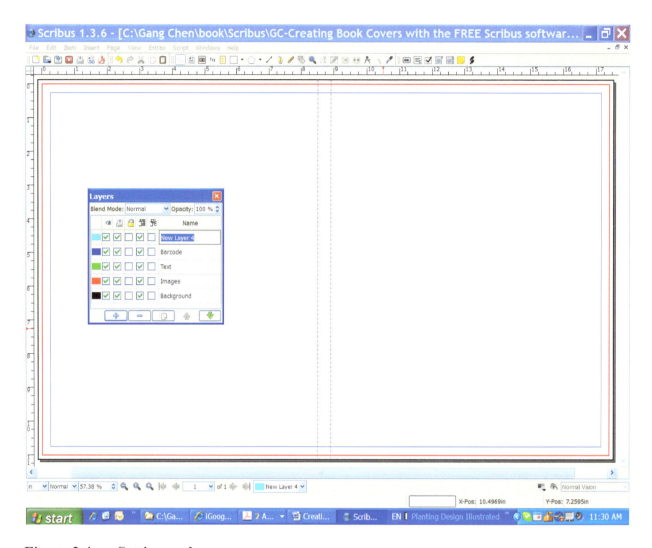

Figure 2.4   Setting up layers

Click the "x" sign to close the pop-up window.

E. Importing the EPS cover template and extracting the barcode
1. **Make sure you are on the "Template" layer.**
    Windows > Layers
    Highlight and select the "**Template**" layer.

2. **Importing the Cover Template (figure 2.5)**
    File > Import > Get Vector File
    A pop-up window will appear. Go to the directory where you saved the cover template, locate and highlight the EPS cover template (9780984374144-Perfect.eps), click OK, and then **click anywhere on the upper left-hand corner** of the screen to import the file.

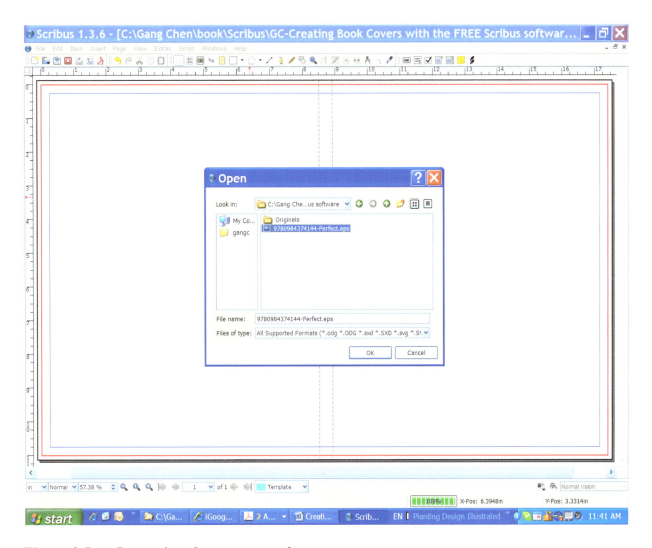

Figure 2.5   Importing the cover template

Your computer screen should look like figure 2.6.

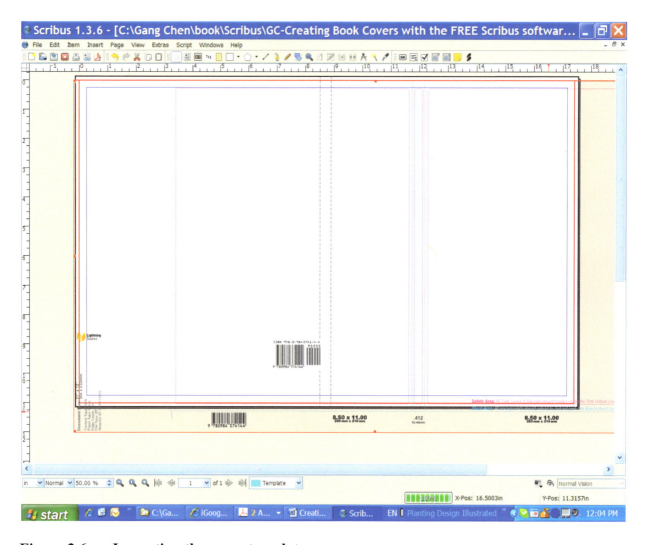

Figure 2.6   Importing the cover template

3. **Moving the EPS cover template to the correct location**
    Currently, your layers may be locked. You can NOT add or modify anything on a layer if it is locked.

    a. Make sure the "Template" layer is unlocked:
        1) **Windows > Layers**
            The box under the lock symbol should be <u>**unchecked**</u> for the "Template" layer.

        2) Also confirm that the box under the lock symbol is **checked** for the "Background" layer and all other layers. You want to lock the "Background" layer so that the book cover you drew earlier will NOT be accidentally moved.

    b. Click on the **edge** of the EPS cover template graphic and move it until the **blue** dotted outline overlaps with the white bleed edge of the cover template you drew earlier. To help you accurately position the template, change the **zoom percentage** on the **lower**

left-hand corner of the screen to 200%. This will enlarge the graphic. You can zoom in and out to check that all the corners align.

View > Fit to Width
Your screen should look like figure 2.7.

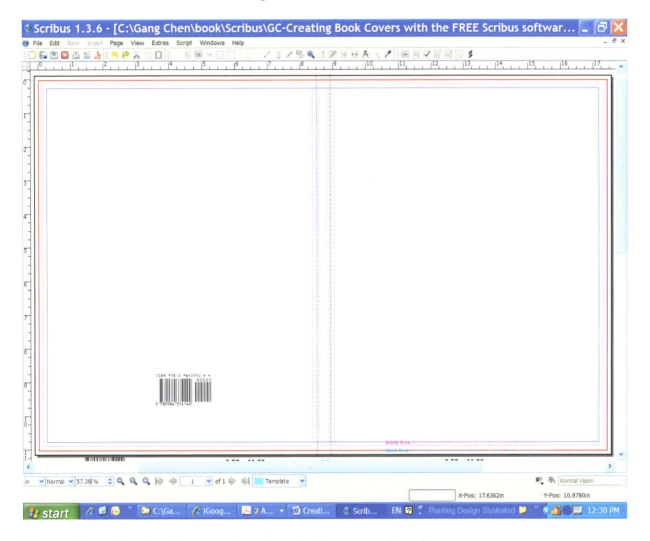

Figure 2.7   Moving the cover template to the correct location

4. **Extracting the barcode from the EPS cover template**
a. **Ungroup the EPS cover template.**
   Right click on the edge of the EPS cover template graphic.
   Item > Ungroup

b. **Send the barcode and the related ISBN to the "Barcode" layer.**
   1) Click on the **barcode** graphic to **highlight**, and then right click.
   Send to Layer > Barcode

   2) Repeat the procedure of step one for the **related ISBN** above the barcode.

3) **Repeat** the procedure of step one for the **white rectangle shape** around the barcode.

Make sure you send **ALL three items** to the "Barcode" layer.

Lock the "Template" layer AFTER you extract these items from the EPS cover template.

You'll notice the **barcode**, the **related ISBN**, and the **white rectangle shape** around the barcode **disappear** after you change them to the "Barcode" layer. This is because the "Barcode" layer is located below the "Template" layer, and is blocked by it.

c. **Moving the barcode layer up** to see the barcode again
Windows > Layers
Click and highlight the "Barcode" layer, and then click the up arrow at the bottom of the pop-up window to move the "Barcode" layer up above the "Template" layer.

The barcode should be visible again. Click the "x" sign to close the pop-up window.

d. **Moving the barcode**
You may need to move the barcode later to avoid a conflict with the text or author's photo. If this is the case, you will want to make sure you move **ALL three items** of the barcode together. The easiest way to do this is to group the elements as one "entity."

You can group the barcode elements as one "entity" by following these directions:
1) **Windows > Layers**
Make sure you highlight the "Barcode" layer.
2) **Edit > Select All**
This command means that ALL items on the **current** "Barcode" layer will be selected. This is why we sent all three barcode elements to the same layer.
3) **Item > Group**

F. **Another way to obtain the barcode from the cover template**
1) Open the PDF file of the cover template.
2) Save the PDF as a JPEG (Barcode.jpg).
3) Use PhotoScape to open the file (Barcode.jpg), and crop only the rectangle area of the barcode.
4) Save the file at the highest resolution possible.
5) You can insert the barcode as a JPEG image later. The **typical size of a Barcode is 1.8" wide x 1" high**. Use these dimensions to insert an accurate **image frame** of the barcode.
6) See discussions under **Adding the images** in chapter three (page 35) for instruction on how to insert the file (Barcode.jpg).

G. Setting colors for the cover

1. RGB or CMYK?

There are two basic categories of colors: RGB and CMYK.

RGB stands for Red, Green, and Blue **light**. These three colors can be combined to create an unlimited number of different color lights. RGB is used primarily for display on light emitting devices such as computer and TV screens. If Red, Green, and Blue light are mixed at 100% strength, they will produce **white** light.

CMYK stands for Cyan, Magenta, Yellow, and Black **pigment**. These four colors can be combined to create an unlimited number of different pigment colors. CMYK is used for color printing. A printed surface will **absorb** or **reflect light** from an **external light source**. If you combine Cyan, Magenta, and Yellow **pigment** on a printed surface at 100% strength, they should produce a **Black** color in theory. In reality, it is very hard to achieve a pure Black color by mixing the Cyan, Magenta, and Yellow pigment. So, black pigment is added.

Therefore, the colors you see on your screen will NOT be the same as the printed copy.

If you reach most of your audience via printed books, you should use CMYK colors. If you reach most of your audience via eBooks, you should use RGB colors.

We chose to use CMYK for our example.

a. Edit > Color

Highlight the blue color on the pop-up window, click Edit, and then select <u>CMYK</u> under **Color Model** (figure 2.8).

b. Repeat step "a" for ALL existing colors listed. When you have finished, click **OK** (NOT the "x" sign) to save your CMYK preferences and close the pop-up window.

Note:
If you click the "x" sign the pop-up window will close, but your preferences will not be saved.

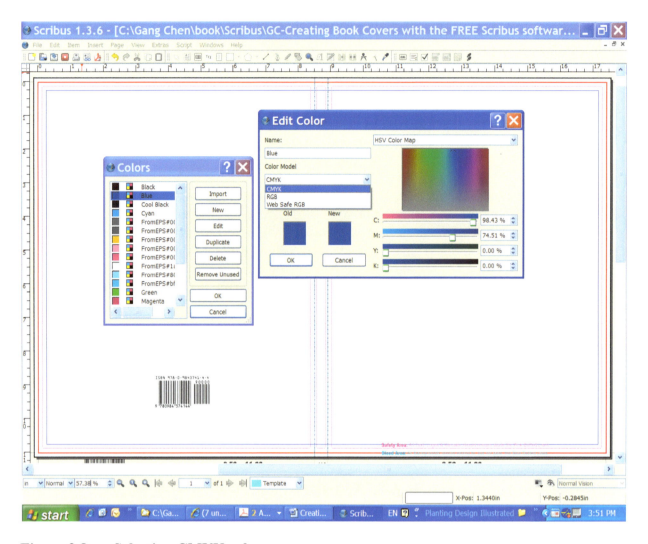

Figure 2.8   Selecting CMYK colors

2. **Creating your own colors**
    We are going to create a new dark green color to use as the background color for our book cover.
    Edit > Color
    On the pop-up window select **New**, and type in **"Dark Green"** under **Name**: to replace "New Color". Click on the color spectrum and select a nice green color. You can adjust the slide bar for C, M, Y, and K under the color spectrum to fine turn the color (figure 2.9).

    Click both the **OK** buttons (NOT the "x" sign) to close the pop-up windows. You can ALWAYS come back and edit the color later if you need to.

    Note:
    Go to art events in museums and galleries to refine your color sensibility. Through experience you will be able to recognize good colors, and then use them for your book covers or any other artistic creation. This is a life-long artistic training.

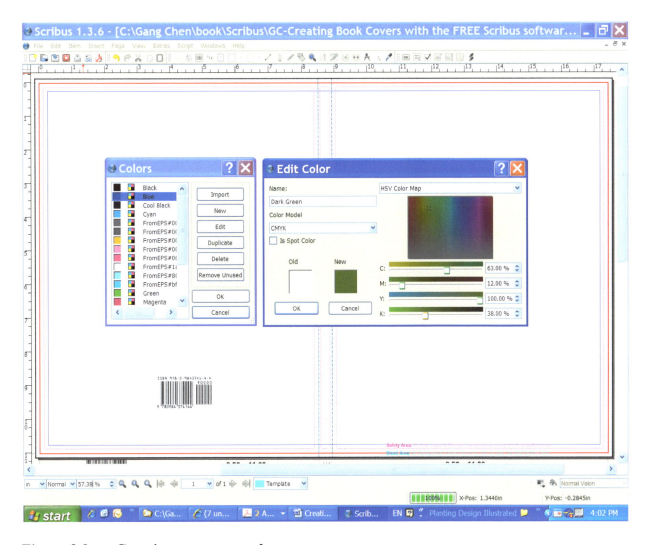

Figure 2.9   Creating your own colors

H. Adding background color to the cover

1. Shape or image frame?

We can use shape or image frame to add background color for our book cover. We prefer to use image frame because the color can be turned on and off as needed.

2. Going to the correct layer

Windows > Layers

Select the "Background Color" layer and use the down arrow to move this layer to the bottom of the list.

Click the "x" sign to close the pop-up window.

3. Adding the background color

a. Insert > Insert Image Frame
Click anywhere on the upper left-hand corner and then click anywhere on the lower right-hand corner, to draw a rectangular **image frame**.

b. Calculating the book cover design width and height:

Paper width = book cover **design** width = left bleed area width + **final** book cover **width** + right bleed area width = 0.125" (left bleed area width) + 8.5" (back cover width) + 0.412" (spine width) + 8.5" (front cover width) + 0.125" (right bleed area width) = **17.662"**

Paper height = book cover **design** height = 0.125" (top bleed area width) + 11" (**final** book cover **height**) + 0.125" (bottom bleed area width) = **11.25"**

c. Adjusting the image frame to <u>accurate</u> dimensions
Right click on the **edge** of the image frame and a pop-up window will appear.

Select Properties and fill in the following information (figure 2.10):
X-pos: <u>-0.125"</u> (**Note:** *This is to locate the top left-hand corner of the bleed area.*)
Y-pos: <u>-0.125"</u> (**Note:** *This is to locate the top left-hand corner of the bleed area.*)
Width: <u>17.662"</u> (**Note:** *This is the paper **width**; see step "b".*)
Height: <u>11.25"</u> (**Note:** *This is the paper **height**; see step "b".*)

The image frame now matches the exact dimensions of paper size, also known as the book cover design, or the **final** book cover PLUS the 0.125" bleed area on all sides.

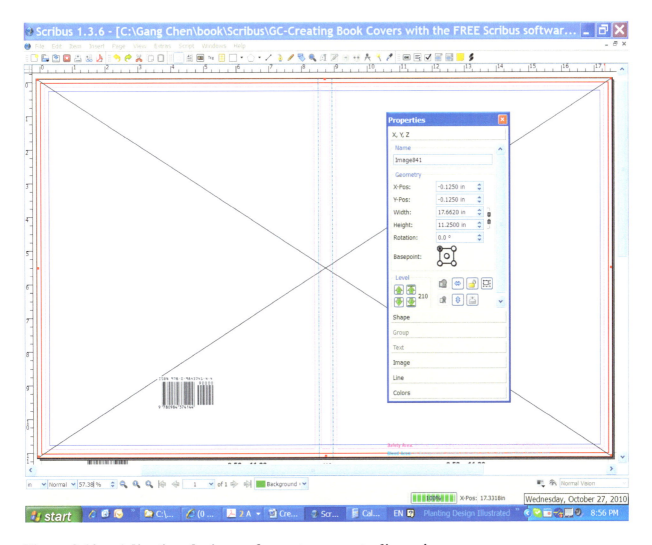

Figure 2.10   Adjusting the image frame to accurate dimensions

d. Adding the background color
Click the **Colors** button at the bottom of the pop-up window, and then scroll down the list to select the Dark Green color. Your screen should look like figure 2.11.

Click the "x" sign to close the pop-up window.

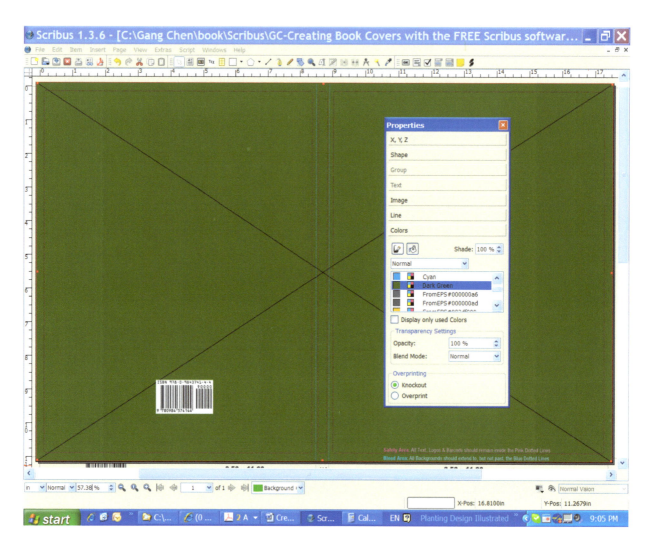

Figure 2.11   Adding the background color

e. **Turning off the background color**
    If you want to turn off the background color of the cover, you can simply make the "Background Color" layer invisible.
    Windows > Layers

    Uncheck the box below the eye symbol for the "Background Color" layer, and the dark green color assigned to this layer will disappear (figure 2.12). If you check this box again, the background color will reappear.

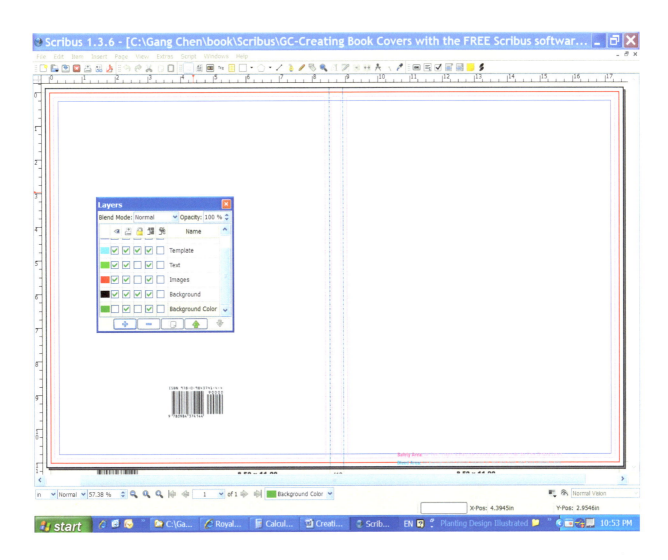

Figure 2.12   Turning off the background color

Chapter Three

Adding the Barcode, Images, Titles, and Text

A. Selecting FREE or affordable images for your book cover

Finding a great image for your book cover is absolutely critical. A breath-taking photo can make your cover stand out from the competition, and look very attractive.

There are many websites that offer FREE public domain images, which you can use for your book cover:
http://www.photos8.com
http://www.pdphoto.org
http://www.public-domain-photos.com
http://www.publicdomainpictures.net
http://www.public-domain-image.com

You will probably need to list the website in your acknowledgements as a condition of use.

There are also other sites that offer very affordable images, which you can use for your book cover:
http://www.istockphoto.com
http://www.bigstockphoto.com
http://www.photos.com

B. Cropping or modifying images

You can use either GIMP or **PhotoScape** to edit images BEFORE you insert them into your book cover. We prefer PhotoScape because it is so easy to resize, crop, and trim JPEG images. You can even crop to round shapes or other free forms.

C. Adding the images

1. Insert > Insert Image Frame

Click on a point close to the right edge of the spine, and then click on another point close to the right edge of the front cover to draw a rectangular **image frame.** The image frame should roughly match the width to height ratio of your cover image, but make sure the height is a little **larger** than the actual photo. This will help you to adjust the image frame to **ACCURATELY** match the book cover width later.

Right click inside the image frame you just created, select **Get Image (figure 3.1)**, and then go the correct folder to select your JPEG cover image.

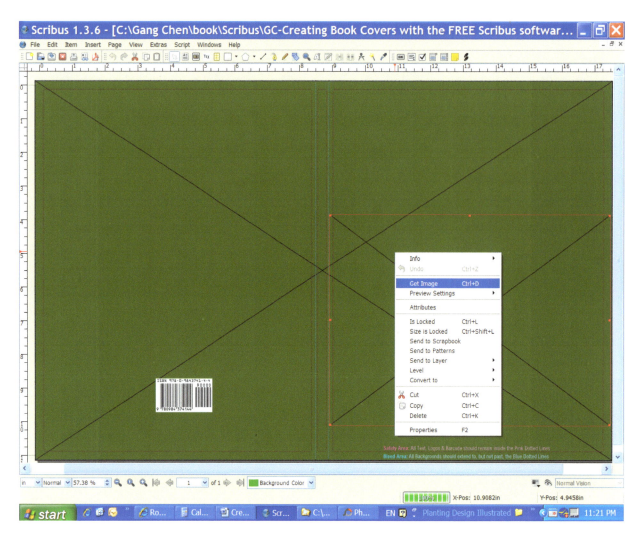

Figure 3.1   Adding the images: Get Image

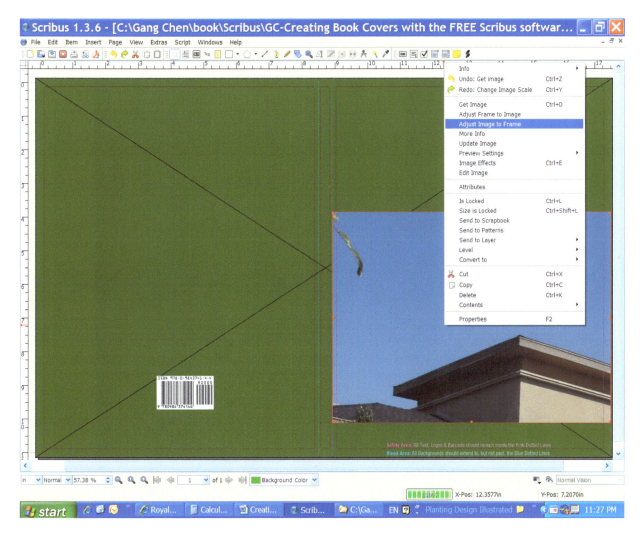

Figure 3.2   Adding the images: Adjust Image to Frame

As you can see from figure 3.2, the image is too large at this point and is out of proportion.

To adjust, **right click** inside the image frame and select **Adjust Image to Frame**. The image should now look like figure 3.3.

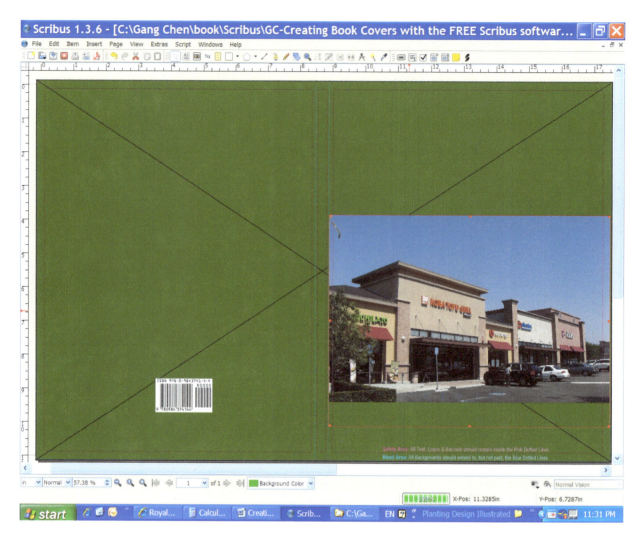

Figure 3.3 **Adding the images: The image looks much better after the "Adjust Image to Frame" command**

2. **Calculating the ACCURATE location of the spine edge and EXACT image width for full bleed images**

   In this example, we want our cover image to bleed from left to right. So, it is **absolutely critical** for the left edge of our cover image to align with the right edge of the book spine

   a. **Calculating the ACCURATE location of the right spine edge**
      X-Pos: = the horizontal distance from the upper left-hand corner of the final book cover to the right spine edge = 8.5" (back cover width) + 0.412" (spine width) = <u>8.912 in</u>

      Width: = image width = 8.5" (front cover width) + 0.125" (bleed area width) = <u>8.625 in</u>

   b. **Right click** inside the image frame, select **Properties,** and then fill in the information from step "a" (figure 3.4).

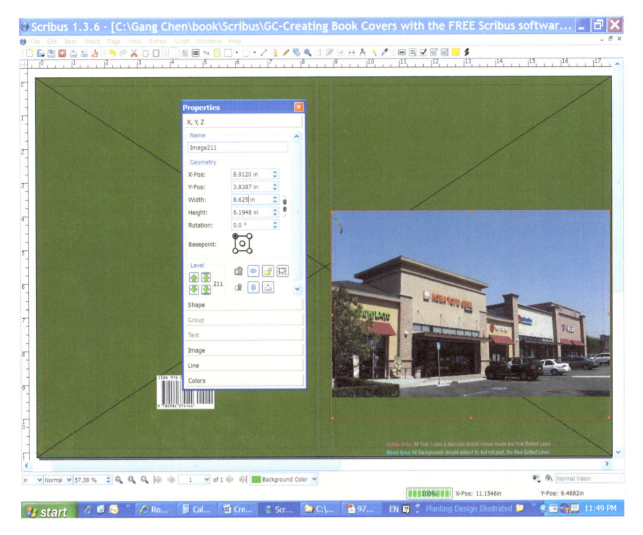

Figure 3.4   **Aligning the image with spine edge ACCURATELY and obtaining the EXACT image width**

Note:
1) **The use of Level**: *See **Level** on the pop-up window of figure 3.4. Level is another feature that you can use to control the display order of text, logos, images, barcodes, etc. The items on a higher level will cover the items on a lower level. We suggest you use **layers** to control the display order **first**, and then use **levels**. A layer has some advantage over a level because you can make a layer completely invisible, and you can organize many similar items on a layer.*
2) **Locking an element**: *See the lock symbol in the middle of the pop-up window of figure 3.4. It is currently unlocked. If you click the button, it will become locked. **Locking an element** (text frame, image frame, etc.) after you finish positioning and adjusting is a good idea **to prevent unintentional moving or modification**.*

c. Click on the bottom of the image **frame** edge, pull it up to match the image edge, and then click the "x" sign to close the pop up window.

D. Adding text

Windows > Layers

Select the "**Text**" layer as the current layer BEFORE you start to work on your text.

1. Selecting the types of fonts and font sizes

You can choose ANY font you wish for a book cover. However, try NOT to use too many types, or your cover may look busy and unorganized.

We suggest using only one or two types of fonts for your cover and prefer **Arial** and **Times New Romans**. Arial is ideal for the title and subtitle due to its clarity. It even reads well when the font size is small such as in a thumbnail graphic on Amazon. Times New Roman is classic and very standard in the publishing industry. Since we generally use this font for the interior content of our books, we sometimes like to include some on the exterior cover as well. Of course, you can do some experimentation and select your own fonts. Just make sure you limit yourself to no more than two types for a book cover.

Since **1" = 72 DTP points** (desktop publishing points), we suggest the following **font styles and sizes** to make your book cover look nice on Amazon and also when people hold the book in hand.

Front Cover:
Title = **Bold** style of **Arial** at **48** points (48/72 = 0.67")
Subtitle = Bold style of Arial at **15** points
Author name = Bold style of Arial at **22** points

Back Cover:
Genre = Regular style **Times New Romans** at **12** points
Title = **Bold** style of **Times New Romans** at **12 to 16** points
Synopsis and Author Bio = Regular style of **Times New Romans** at **10 to 12** points

Note:
You can also use the bold style of Arial on the back cover for external consistency.

Let's start by adding the title to the front cover.

2. Calculating the ACCURATE location of the left text frame edge and the text frame width

We want to leave a 0.25" margin around the final front book cover.

X-Pos: = the horizontal distance from the upper left-hand corner of the final book cover to the right text frame edge = 8.5" (back cover width) + 0.412" (spine width) + 0.25" (left front cover margin) = 9.162 in

Width: = text frame width = 8.5" (front cover width) - 0.25" (left front cover margin) - 0.25" (right front cover margin) = 8.0 in

3. **Adding text accurately and drawing an ACCURATE text frame for the title**
    Insert > Insert Text Frame
    Click on a point close to the right edge of the spine, and click on another point close to the right edge of the front cover to draw a rectangular **text** frame that roughly covers the area above your image.

    Right click inside the text frame, select **Properties,** and fill in the information from step two.

4. **Adding and centering the text**
    Click inside the text frame, and type in the title and subtitles of the book:

    Building Construction

    Project Management, Construction Administration, Drawings, Specs, Detailing Tips, Schedules, Checklists, and Secrets Others Don't Tell You (Architectural Practice Simplified, 2nd edition)

    Highlight the title, **right click,** and a pop-up window will appear. **Change the font for the Title** to **Bold** style **Arial** at **48** points. Click the **center-justify button** below 48.00 pt to make the title centered (figure 3.5).

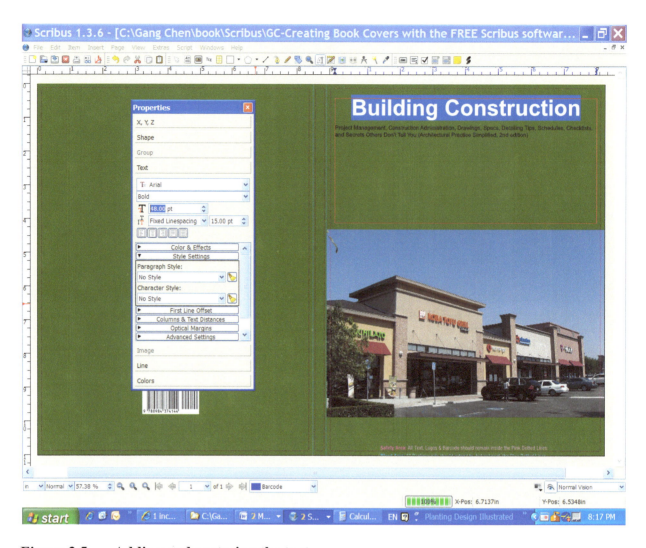

Figure 3.5   Adding and centering the text

5. **Changing the text color**
    Right click inside the text frame.
    Properties > Text > Color and Effects
    Select the **yellow** color from the **pull-down window** below **Color and Effects**. The title text color will change from black to yellow (figure 3.6).

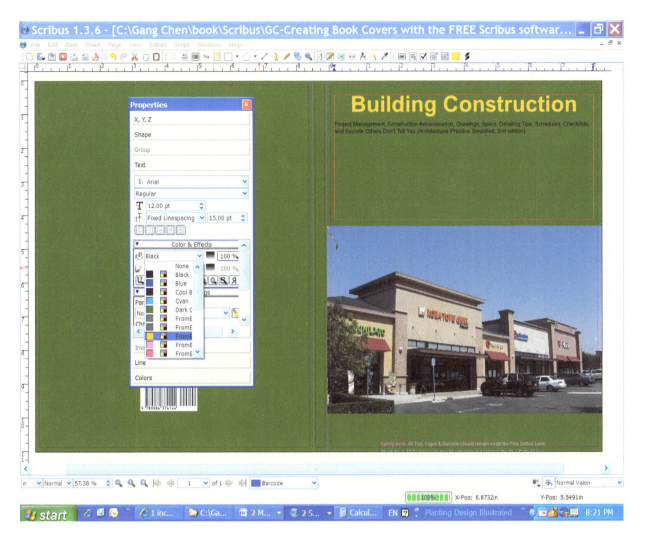

Figure 3.6   Changing the text color

6. **Rotating and aligning the spine text**
1) **Adding a narrow rectangular text frame for the spine text**
   Insert > Insert Text Frame
   Click on a point in the upper left-hand corner of the front cover, and then click on another point close to the right edge of the front cover to draw a **narrow** horizontal rectangular **text** frame (figure 3.7).

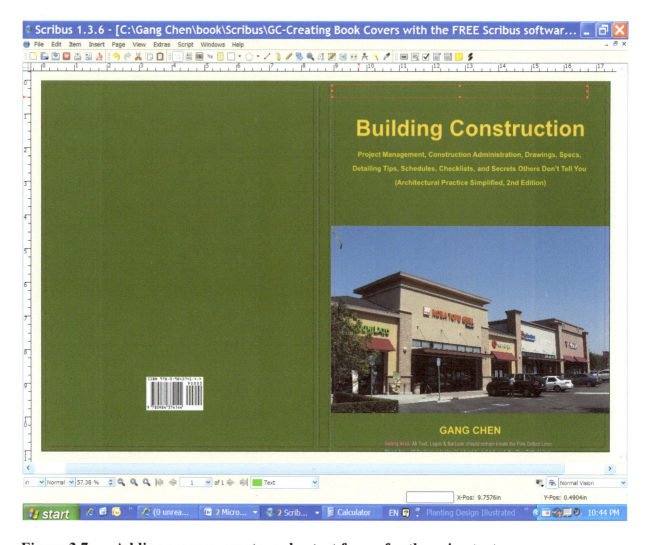

Figure 3.7   Adding a narrow rectangular text frame for the spine text

2) **Calculating the ACCURATE location of the left spine text frame edge and the EXACT text frame width and height**
The spine text will have to be rotated 90 degrees clockwise so that when the book is placed on a desk, with the front cover facing up, the text can be read.

Before rotating 90 degrees, determine the **ACCURATE** location of the **upper left-hand corner** of the spine text frame edge and the **EXACT** text frame width and height:

X-Pos: = 8.5" (final back cover width) + 0.412" (spine width) - 0.0625" (the 1/16" wide right spine margin) = 8.8495 in

Y-Pos: = 0.125 in (the top spine margin)

Width: = the spine text frame = 11" (final book height) – 0.125" (the top spine margin) - 0.125" (the bottom spine margin) = 10.75 in

Height: = the spine text frame = 0.412" (spine width) - 0.0625" (the 1/16" wide left spine margin) - 0.0625" (the 1/16" wide right spine margin) = <u>0.287 in</u> = the maximum font size for spine text

0.287" x 72 = 20.66 (DTP points) = about **20 DTP points**

3) **Right click** inside the text frame, select **Properties** and fill in the information from step two. Your screen should look like figure 3.8. Click **Text**, and then change the font size to **bold** style 18 points (a little smaller than the maximum font size for spine text). If you use the maximum font size the text may NOT show up due to the text box height limitation. Click the "x" sign to close the pop-up window.

The information you fill in under **Rotation**: is <u>270.0^0</u>. Of course, you can rotate the text to ANY angle you want for other profession presentations.

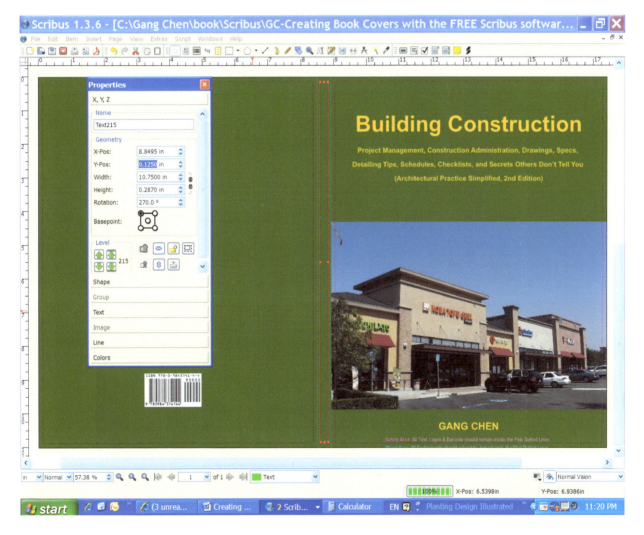

Figure 3.8   **Locating the spine text frame accurately and showing the EXACT spine text frame width and height**

4) Right click inside the text frame. Select **Edit Text** and the **Story Editor** window will pop up. Fill in the following information and adjust the distance between the words to make them look good (figure 3.9):

BUILDING CONSTRUCTION   Gang Chen   ArchiteG.com

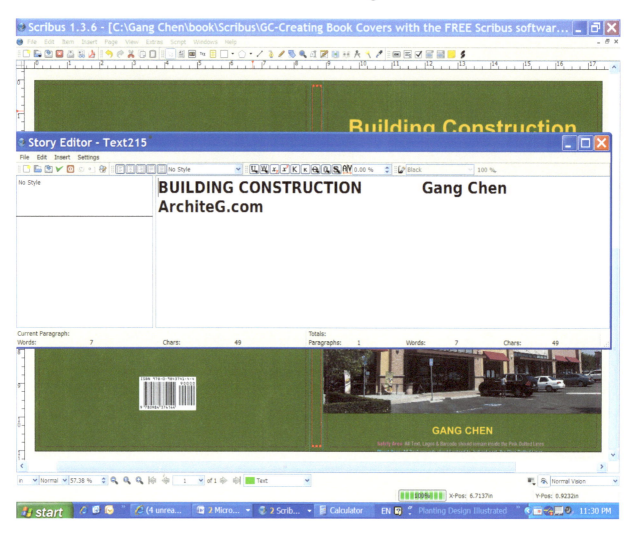

Figure 3.9   Using Story Editor to add spine text

Click the **center-justify button** on the top menu of the Story Editor to make the spine text centered.

Click the "x" sign to close the Story Editor window, and the spine text will show up in black (figure 3.10).

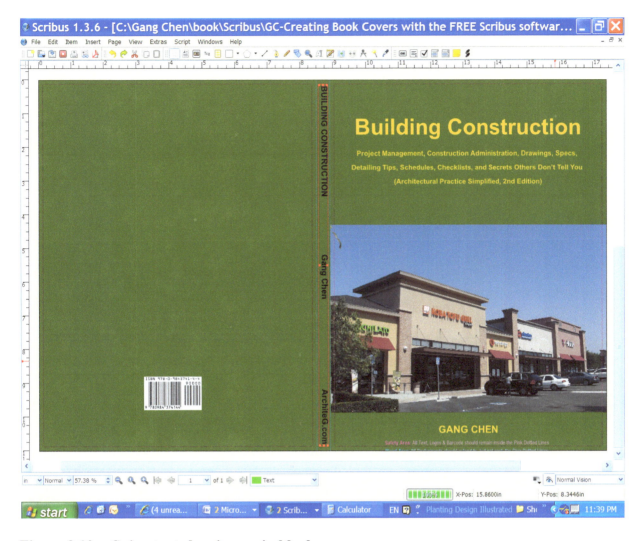

Figure 3.10   Spine text showing up in black

5) **Changing the spine text color**
    Right click inside the spine text frame.
    Properties > Text > Color and Effects
    Select the **yellow** color from the **pull-down window** below **Color and Effects**. The spine text color will change from black to yellow.

6) **Using a guide at the spine center to fine-tune the spine text location**
    X-Pos: for the **spine** centerline = 8.5" (final back cover width) + 0.412"/2 (half of the spine width) = <u>8.706"</u>

    Page > Manage Guides > Single
    Add a guide at 8.706" location.

    Window > Layers
    Highlight the "Background Color" layer and **uncheck** the box under the eye symbol to make the "Background Color" layer invisible. Your screen should look like figure 3.11.

Use the **zoom button** on the lower left-hand corner of the screen to Zoom in at 200%. Manually move the spine text so that the **center** of letter **"O"** in the word "CONSTRUCTION" aligns with the **spine centerline guide**.

Important Note:
*You need to go to **Window > Layers**, **highlight** and select the "Text" layer as your current layer in order to move the text frame.*

Window > Layers
Highlight the "Background Color" layer, and **check** the box under the eye symbol to make the "Background Color" layer visible again.

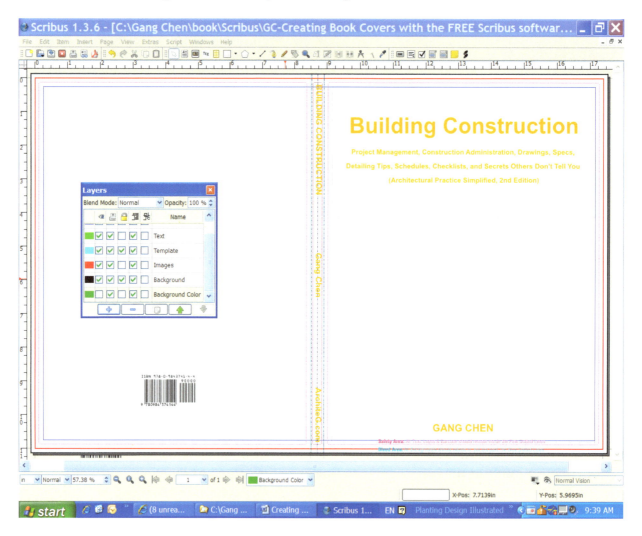

Figure 3.11   Making the "Background Color" layer invisible to fine-tune the spine text location with the centerline guide

7. **Repeat steps one through five, to add or modify the remaining text**: author name on the front cover, synopsis, and author bio on the back cover, etc. You can add extra spaces between the text lines to make the cover look better.

We decided to change both the title and subtitle to ALL upper case letters.

8. Important Notes:

1) You may want to use two or three text frame boxes for the back cover, so that there is room for the author's photo and barcode.

2) Make sure you use **X-Pos:** and **Width:** to **align** the different text frames on the front and back cover. Your screen should look like figure 3.12 after you add all the remaining text, the author's photo, and move the barcode. Figure 3.12 shows the finished book cover with the "Template" layer visible. Confirm that the text, logo, images, and barcode are within the safety area.

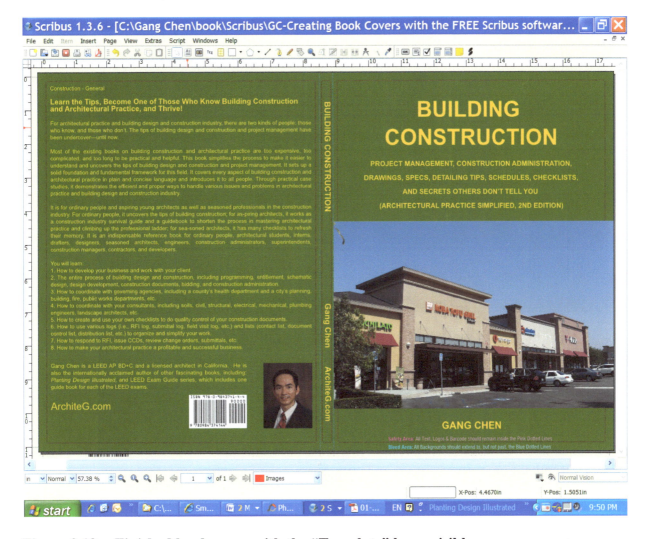

Figure 3.12   Finished book cover with the "Template" layer visible

Windows > Layers

Select the "Template" layer and **uncheck** the box under the **eye symbol AND printer symbol** to make the "Template" layer **invisible AND unprintable**. Your computer screen should look like figure 3.13.

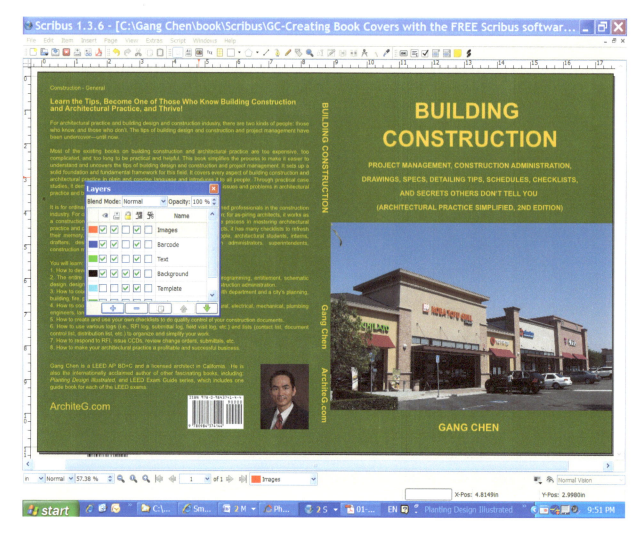

Figure 3.13   Finished book cover with the "Template" layer invisible AND unprintable

E. Exporting to a PDF file

1. **File > Export > Save as PDF**

    A pop-up window appears. Click the arrow next to **Compatibility** under **Files Option** and select **PDF1.3 (Acrobat 4)**. Then click the **Color** tab and the arrow next to **Output Intended For;** select **Printer**, and click **Save**.

    This creates a **PDF file** that can be submitted to your printer for covers of books with **black and white** interiors. Sometimes you may get an error message when you perform this task. If so, simply click **Ignore Errors** and normally you still end up with a good PDF file (figure 3.14).

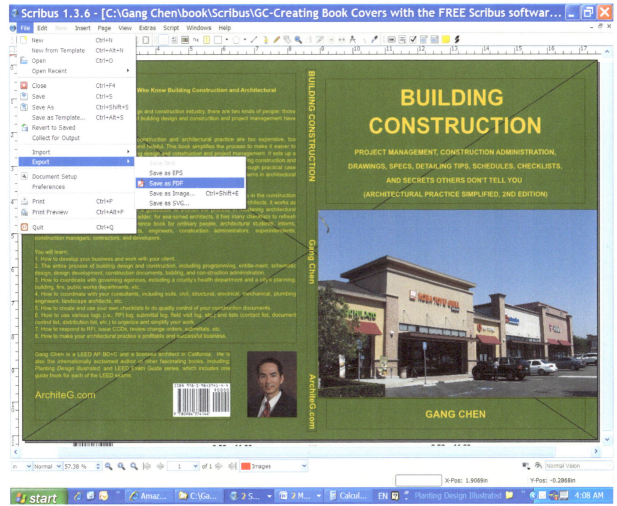

Figure 3.14   Exporting to a PDF file

Note:
*The previous step is the easiest way to create a PDF for book covers with a **black and white** interior, but is probably NOT acceptable for covers with a **color** interior. Most book printers require PDF/x-1a:2001 compatible files for covers with **color** interiors. See Section H (page 54) for instructions on how to create PDF/x-1a:2001 compatible files.*

2. **Open the PDF file you just made to double-check (figure 3.15)**
   a. **Confirm** that the cover looks good. Sometimes you may have forgotten to make certain layers invisible AND unprintable, and you might have some extra lines that you do NOT want on the PDF.
   b. Double check the PDF file to make sure that **all fonts are embedded**.
      File > Properties
      Click **Fonts** in the pop-up window and you should see (Embedded) or (Embedded Subset) next to each of the fonts listed. If your fonts are not fully embedded, you need to go back and change them. If you don't, you run the risk of the book printer using substitute fonts, which may not appear the same as the fonts on your computer. Most book printers are NOT very strict on this requirement for book covers with black and

white interiors. I know some people who have submitted files without 100% embedded fonts, which were still accepted and turned out looking good. However, it's always good to be cautious.

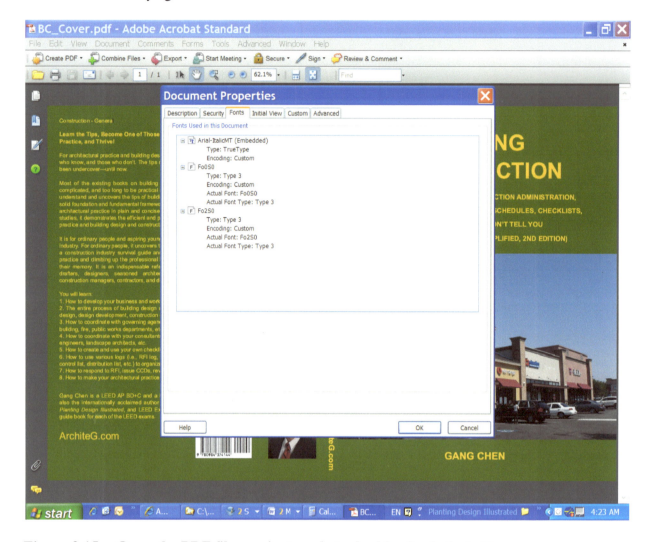

Figure 3.15   Open the PDF file you just made to double-check that all unwanted layers are invisible AND unprintable, and all fonts are embedded

F. Making revisions (figure 3.16)

You can always go back at any time to modify the book cover. For example, if you want to change the front cover image, you can easily replace it.

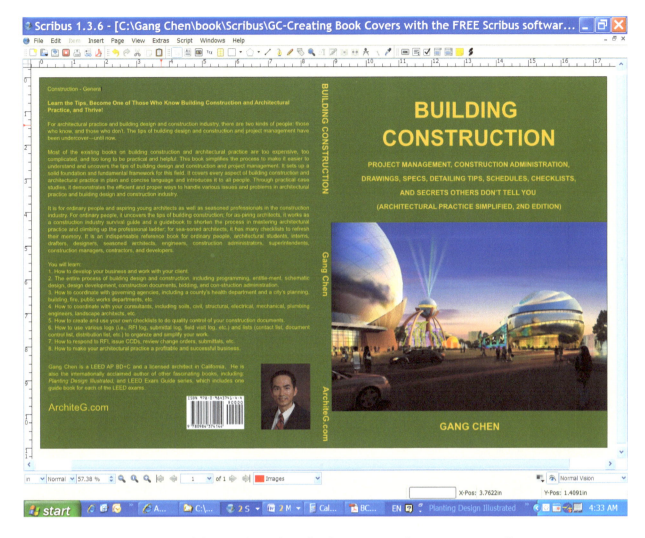

Figure 3.16   Making revisions: changing the front cover image or any other element

G. Modifying for NO bleeds eBook covers

Covers on eBooks have **NO bleed area**. To change your book cover, follow these directions:

1. Save the Scribus file you created as **another** file first (**VERY important**). Name the new file (ISBN_eBook_Cov.sla).
2. **Edit > Color > Edit**
    Set all colors to RGB colors.
3. **File > Document Setup > Bleeds**
    Change all bleeds from 0.125" to 0".
4. **File > Export > Save as PDF**
    A pop-up window appears. Click the arrow next to **Compatibility** under **Files Option** and select **PDF1.3 (Acrobat 4)**. Then click the **Color** tab and the arrow next to **Output Intended For;** select **Screen/Web**, and then **Save**. This creates a **PDF file**.
5. Save the PDF file as a high resolution JPEG file.

6. Use PhotoScape to crop the JPEG file, and save as two files (ISBN_front_cov.jpg and ISBN_back_cov.jpg).

H. Special requirements for book covers with COLOR interiors

Most fiction and non-fiction books have black and white interiors, but sometimes you may need to design covers for books with a COLOR interior.

Scribus is excellent for designing book covers that have black and white interiors, but it requires a little extra work and the help of commercial software for those with **color** interiors. The following are special requirements for book covers that have **color** on the interior:

1. The bleed area is normally 1/4" instead of a 1/8" for all edges.

2. The book printer will probably insist on having your book cover placed on their specific template for books that have interior COLOR. For example, the book printer's template size for this particular book is 20.5 x 14.33" and they require you to upload a PDF file this same size, even though the final book cover size is actually 17.412" x 11". Even though the book printer has exactly the same requirement for covers with black and white interior books, they NEVER enforce this requirement.

It takes some extra effort to do this, and you can meet this requirement in one of the following ways:

a. You can set up the entire sheet based on the book printer's cover template from the very beginning, and not have to worry about it later. The disadvantage of this approach is that the math used to calculate the exact location of the text box, image frame, etc. is more complicated. The steps to do this are very similar to our previous instructions, only the dimensions will be different.

b. You can set up the book cover per our previous instructions, and then move all book cover elements to match the cover template.

Before you make these revisions, save your file as another file with a different name because the changes are very hard to undo later. The trick to moving the cover elements is to **move them all at once instead of individually.** First **send** all the elements to the same layer. Click **Select All**, **group** them as one item, and move them together to match the cover template. You may have to use **level** to move the display order of the cover elements to make sure they display properly with image or text above background color, etc.

c. You can export your book cover as an EPS file, and then create a new file with a page size of 20.5" width x 14.33" height. Insert the original book printer's EPS template file and position so that its edges match the page edges of the new document. Then insert your recently exported EPS file and align the cover with the book template outline.

Some points that you should be aware of are as follows:

1) When you export an EPS file, ONLY the elements included in the area defined by the page **Width** and **Height** will show up; the elements defined by the **Bleeds** command of Scribus will NOT show up.
2) You may need to add an extra image frame with the background color to cover the blank bleed area around the book cover on the EPS file.

3. Most book printers require **PDF/x-1a:2001** compatible files for the cover and content of books with interior **color**, and are VERY strict about this. Some printers also require **PDF/x-1a:2001** compatible files for the cover and content of books with a **black and white interior**, but do NOT really enforce this requirement.

4. After completing the previous steps, you need to export your book cover as an **EPS** file from Scribus. Use Adobe Acrobat Distiller 8 Professional or newer to open the EPS file and convert it to a **PDF/x-1a:2001** compatible file. This program is often sold as part of Adobe Acrobat 8 Professional or newer. The software bundle costs about $120 for a student & teacher version, and $450 for a regular version. You must contact and prove to Adobe that you are actually a full-time student or teacher BEFORE you can get the code to activate the student version. If you have a child in primary or secondary school or college, you can buy the student version for her/him. You can download the free trial version of the software, but this version does NOT seem to have all the functions and expires after 30 days. Another option is to go to FedEx Kinko's and spend $0.45 per minute to rent a computer with Adobe Acrobat Distiller 8 Professional or newer. It will probably cost you $10.00 in total rental fees to convert and check the PDF file against your book printer's PDF requirements. See detailed discussions on the requirements later.

Exporting your book cover as an EPS file from Scribus
File > Export > Save as EPS
A pop-up window appears. Click OK and an EPS file will be made. Sometimes you may have an error message. If so, click **Ignore Errors**.

Opening the EPS file and converting it to a PDF/x-1a:2001 compatible file
a. After you install Adobe Acrobat Distiller 8 Professional or newer, an Acrobat Distiller Pro icon appears on your desktop. Double-click on the icon to open.

b. Under **Default Settings**, click on the **down arrow**, and select **PDF/X-1a: 2001**.
   Settings > Edit Adobe PDF Settings
   A pop-up window appears. Fill in the Page **Width** and **Height**. For our example, fill in the following (figure 3.17):
   Width: <u>17.677"</u>
   Height: <u>11.50"</u>

   Click **OK**, and save the **setting** as **PDF/X-1a 2001**.

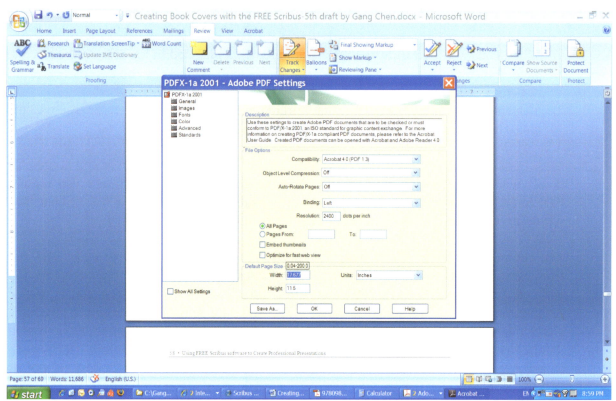

Figure 3.17   Edit Adobe PDF Settings for Adobe Acrobat Distiller 8 Professional or newer

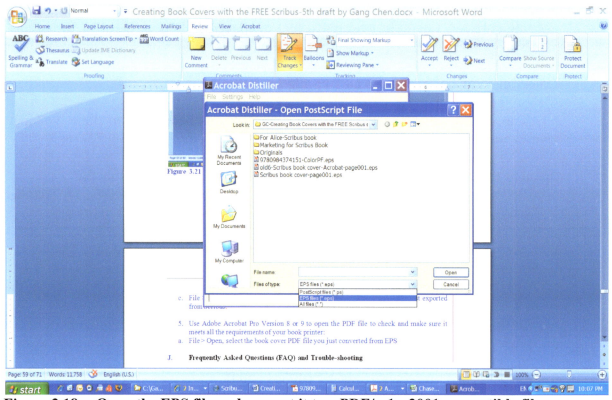

Figure 3.18   Open the EPS file and convert it to a PDF/x-1a:2001 compatible file

c. **File > Open**

A pop-up window appears. Click on the down arrow next to **Files of type:** and select **EPS files (*.eps)**. Go to the proper folder and select the EPS file you just exported from Scribus (figure 3.18). Another pop-up window appears. Click **Save** to create a PDF file.

Another way to create a PDF/X-1a:2001 compliant file:

Export the file from Scribus as a PDF file. Using Adobe Acrobat 8 Pro or newer, save it as an EPS file and then use Adobe Acrobat Distiller 8 or newer to open the EPS file and save it as a PDF/X-1a:2001 compliant file.

This is probably the easiest and best way to create a PDF/X-1a:2001 compliant file. This method can retain more information from Scribus, and the fonts look better than if exporting an EPS file from Scribus directly.

However, Scribus is constantly improving and the upcoming Scribus 1.5.0 or newer will probably have the capacity to generate PDF/X-1a:2001 compliant files directly.

5. Use Adobe Acrobat Pro Version 8 or 9 to open the PDF file and check to make sure it meets all your book printer's requirements.

a. **File > Open**

Select the PDF file you just converted from EPS.

b. Check Document Dimensions.

File > Properties

Click the **Description** Tab to check the **Page Size** of your PDF file.

c. Check Font Embedding by clicking the **Fonts** Tab. You should see (Embedded) or (Embedded Subset) next to each of the fonts.

d. Check Color Mode and Resolution.

Advanced > Print Production > Preflight

e. Check Spot Colors.

Advanced > Print Production > Output Preview

From your book printer's website, you can probably obtain detailed instructions on how to check a PDF file BEFORE submittal.

I. **Using Scribus software to create other professional presentations**

Book covers are one of the most complicated forms of professional presentations, which is why we chose a book cover case study to walk you through the entire process of creating professional presentations with Scribus.

You should now be able to use the knowledge that you have gained through this book to create magazine covers, graphic designs, posters, newsletters, renderings, and more.

J. Frequently Asked Questions (FAQ) and Trouble-shooting

1. **When I export to a PDF file, only part of the book cover showed up. What happened? How can I fix it?**

   Answer: You probably accidentally set up the page at the wrong size. For example, the book cover we created for *Building Construction* should be 17.412" x 11" PLUS the 1/8" wide bleed area on all four sides. If you accidentally set up the page at 17" x 11", only the 17" x 11" portion of the book cover will show up. Submitting the wrong sized PDF file to your book printer will result in rejection and additional fees to submit again.

   To check or correct this problem do the following:
   a. **File > Document Setup**
      Make sure your page size is set at 17.412" x 11". Click **Bleeds** and make sure the **bleed area width** is set at <u>0.125 in</u> for all four sides.

   b. **Page > Manage Page Properties**
      Make sure your page size is set at 17.412" x 11".

2. **How do I insert a round or irregular shaped front cover image?**

   Answer: You need to create a shape, convert it to an image frame, and insert an image.
   a. **Insert > Insert Shape >Default Shapes**
      Select the shape you need. In the case of figure 0.2, the shape is a circle. To create an oval, **right click** on the circle, select **Properties**, and change the **width and height**.

   b. **Exit** the pop-up window and **right click** on the shape.
      Properties > Convert to > Image frame

   c. **Exit** the pop-up window, **right click** on the Image frame you just converted, and select **Get Image**.

   d. **Exit** the pop-up window, **right click** on the Image frame, and select **Adjust Image to Frame**.

3. **What else can I create with Scribus?**

   Answer: Scribus is a very powerful program and there are many other art projects you can create with it. We have introduced only the most fundamental information in this book. Play with the program. If you become an expert, you might even be able to make a living by creating professional presentations with it.

Appendixes

A. List of Figures

B. Annotated Bibliography

Christoph Schäfer & Gregory Pittman. *Scribus: Open-Source Desktop Publishing*. FLES Books Ltd. 2009. This is the official Scribus manual, which contains very detailed information. This text is useful for looking up difficult Scribus operating information and procedures. While thorough, it may contain too much information if you just want to learn how to create a quick book cover design. 452 pages. 8.9 x 5.9 x 1.1 inches.

Back Page Promotion

You may be interested in some other books written by Gang Chen:

1. *Planting Design Illustrated*
   http://GreenExamEducation.webs.com/

2. **LEED Exam Guides series.** See the link below:
   http://www.ArchiteG.com

Note: Other books in the **LEED Exam Guides series** are in the process of being produced. **One book will eventually be produced for each of the LEED exams.** The series include:

LEED AP Exam Guide: Study Materials, Sample Questions, Mock Exam, Building LEED Certification (LEED NC v2.2), and Going Green, **Book 1,** LEED Exam Guides series, LEEDSeries.com (Published on 9/23/2008).

LEED GA EXAM GUIDE: A Must-Have for the LEED Green Associate Exam: Comprehensive Study Materials, Sample Questions, Mock Exam, Green Building LEED Certification, and Sustainability (LEED v3.0), Book 2, LEED Exam Guide series, ArchiteG.com (Published October 28, 2009)

LEED BD&C EXAM GUIDE: A Must-Have for the LEED AP BD+C Exam: Comprehensive Study Materials, Sample Questions, Mock Exam, Green Building Design and Construction, LEED Certification, and Sustainability (LEED v3.0), Book 3, LEED Exam Guide series, ArchiteG.com (Published December 18, 2009)

LEED ID&C EXAM GUIDE: A Must-Have for the LEED AP ID+C Exam: Comprehensive Study Materials, Sample Questions, Mock Exam, Green Interior Design and Construction, LEED Certification, and Sustainability, Book 4, LEED Exam Guide series, ArchiteG.com (Published March 8, 2010)

LEED O&M EXAM GUIDE: A Must-Have for the LEED AP O+M Exam: Comprehensive Study Materials, Sample Questions, Mock Exam, Green Building Operations and Maintenance, LEED Certification, and Sustainability (LEED v3.0), Book 5, LEED Exam Guide series, ArchiteG.com

LEED HOMES EXAM GUIDE: A Must-Have for the LEED AP Homes Exam: Comprehensive Study Materials, Sample Questions, Mock Exam, Green Building LEED Certification, and Sustainability, Book 6, LEED Exam Guide series, ArchiteG.com

LEED ND EXAM GUIDE: A Must-Have for the LEED AP Neighborhood Development Exam: Comprehensive Study Materials, Sample Questions, Mock Exam, Green Building LEED Certification, and Sustainability, Book 7, LEED Exam Guide series, ArchiteG.com

LEED GA MOCK EXAMS: Questions, Answers, and Explanations: A Must-Have for the LEED Green Associate Exam, Green Building LEED Certification, and Sustainability, Book 8, LEED Exam Guide series, ArchiteG.com (Published August 6, 2010)

LEED O&M MOCK EXAMS: *Questions, Answers, and Explanations: A Must-Have for the LEED O&M Exam, Green Building LEED Certification, and Sustainability*, Book 9, LEED Exam Guide series, ArchiteG.com (Published September 28, 2010)

How to order these books:
You can order them at
http://GreenExamEducation.webs.com/

OR
http://www.ArchiteG.com

Following are some detailed descriptions:

LEED Exam Guides series*:* Comprehensive Study Materials, Sample Questions, Mock Exam, Building LEED Certification and Going Green

LEED (Leadership in Energy and Environmental Design) is the most important trend of development, and it is revolutionizing the construction industry. It has gained tremendous momentum and has a profound impact on our environment.

From LEED Exam Guides series, you will learn how to

1. Pass the LEED Green Associate Exam and various LEED AP + exams (each book will help you with a specific LEED exam).

2. Register and certify a building for LEED certification.

3. Understand the intent for each LEED prerequisite and credit.

4. Calculate points for a LEED credit.

5. Identify the responsible party for each prerequisite and credit.

6. Earn extra credit (exemplary performance) for LEED.

7. Implement the local codes and building standards for prerequisites and credit.

8. Receive points for categories not yet clearly defined by USGBC.

There is currently NO official book on the LEED Green Associate Exam, and most of the existing books on LEED and LEED AP are too expensive and too complicated to be practical and helpful. The pocket guides in LEED Exam Guides series fill in the blanks, demystify LEED, and uncover the tips, codes, and jargon for LEED as well as the true meaning of "going green." They will set up a solid foundation and fundamental framework of LEED for you. Each book in the LEED Exam Guides series covers every aspect of one or more specific LEED rating system(s) in plain and concise language and makes this information understandable to all people.

These pocket guides are small and easy to carry around. You can read them whenever you have a few extra minutes. They are indispensable books for all people—administrators; developers;

contractors; architects; landscape architects; civil, mechanical, electrical, and plumbing engineers; interns; drafters; designers; and other design professionals.

Why is the LEED Exam Guides series needed?

A number of books are available that you can use to prepare for the LEED Exams:

1. *USGBC Reference Guides.* You need to select the correct version of the *Reference Guide* for your exam.

   The *USGBC Reference Guides* are comprehensive, but they give too much information. For example, *The LEED 2009 Reference Guide for Green Building Design and Construction (BD&C)* has about 700 oversized pages. Many of the calculations in the books are too detailed for the exam. They are also expensive (approximately $200 each, so most people may not buy them for their personal use, but instead, will seek to share an office copy).

   It is good to read a reference guide from cover to cover if you have the time. The problem is not too many people have time to read the whole reference guide. Even if you do read the whole guide, you may not remember the important issues to pass the LEED exam. You need to reread the material several times before you can remember much of it.

   Reading the reference guide from cover to cover without a guidebook is a difficult and inefficient way of preparing for the LEED AP Exam, because you do NOT know what USGBC and GBCI are looking for in the exam.

2. The USGBC workshops and related handouts are concise, but they do not cover extra credits (exemplary performance). The workshops are expensive, costing approximately $450 each.

3. Various books published by a third party are available on Amazon. However, most of them are not very helpful.

   There are many books on LEED, but not all are useful.

   LEED Exam Guides series will fill in the blanks and become a valuable, reliable source:

   a. They will give you more information for your money. Each of the books in the LEED Exam Guides series has more information than the related USGBC workshops.

   b. They are exam-oriented and more effective than the USGBC reference guides.

   c. They are better than most, if not all, of the other third-party books. They give you comprehensive study materials, sample questions and answers, mock exams and answers, and critical information on building LEED certification and going green. Other third-party books only give you a fraction of the information.

   d. They are comprehensive yet concise. They are small and easy to carry around. You can read them whenever you have a few extra minutes.

   e. They are great timesavers. I have highlighted the important information that you need to understand and MEMORIZE. I also make some acronyms and short sentences to help you easily remember the credit names.

It should take you about 1 or 2 weeks of full-time study to pass each of the LEED exams. I have met people who have spent 40 hours to study and passed the exams.

You can find sample texts and other information on the LEED Exam Guides series in customer discussion sections under each of my book's listing on Amazon.

What others are saying about *LEED GA Exam Guide* (Book 2, LEED Exam Guide series):

"Finally! A comprehensive study tool for LEED GA Prep!

"I took the 1-day Green LEED GA course and walked away with a power point binder printed in very small print—which was missing MUCH of the required information (although I didn't know it at the time). I studied my little heart out and took the test, only to fail it by 1 point. Turns out I did NOT study all the material I needed to in order to pass the test. I found this book, read it, marked it up, retook the test, and passed it with a 95%. Look, we all know the LEED GA exam is new and the resources for study are VERY limited. This one's the VERY best out there right now. I highly recommend it."
—**ConsultantVA**

"Complete overview for the LEED GA exam

"I studied this book for about 3 days and passed the exam … if you are truly interested in learning about the LEED system and green building design, this is a great place to start."
—**K.A. Evans**

"A Wonderful Guide for the LEED GA Exam

"After deciding to take the LEED Green Associate exam, I started to look for the best possible study materials and resources. From what I thought would be a relatively easy task, it turned into a tedious endeavor. I realized that there are vast amounts of third-party guides and handbooks. Since the official sites offer little to no help, it became clear to me that my best chance to succeed and pass this exam would be to find the most comprehensive study guide that would not only teach me the topics, but would also give me a great background and understanding of what LEED actually is. Once I stumbled upon Mr. Chen's book, all my needs were answered. This is a great study guide that will give the reader the most complete view of the LEED exam and all that it entails.

"The book is written in an easy-to-understand language and brings up great examples, tying the material to the real world. The information is presented in a coherent and logical way, which optimizes the learning process and does not go into details that will not be needed for the LEED Green Associate Exam, as many other guides do. This book stays dead on topic and keeps the reader interested in the material.

"I highly recommend this book to anyone that is considering the LEED Green Associate Exam. I learned a great deal from this guide, and I am feeling very confident about my chances for passing my upcoming exam."
—**Pavel Geystrin**

"Easy to read, easy to understand

"I have read through the book once and found it to be the perfect study guide for me. The author does a great job of helping you get into the right frame of mind for the content of the exam. I had started by studying the Green Building Design and Construction reference guide for LEED projects produced by the USGBC. That was the wrong approach, simply too much information with very little retention. At 636 pages in textbook format, it would have been a daunting task to get through it. Gang Chen breaks down the points, helping to minimize the amount of information but maximizing the content I was able to absorb. I plan on going through the book a few more times, and I now believe I have the right information to pass the LEED Green Associate Exam."
—Brian Hochstein

"All in one—LEED GA prep material

"Since the LEED Green Associate exam is a newer addition by USGBC, there is not much information regarding study material for this exam. When I started looking around for material, I got really confused about what material I should buy. This LEED GA guide by Gang Chen is an answer to all my worries! It is a very precise book with lots of information, like how to approach the exam, what to study and what to skip, links to online material, and tips and tricks for passing the exam. It is like the 'one stop shop' for the LEED Green Associate Exam. I think this book can also be a good reference guide for green building professionals. A must-have!"
—SwatiD

"An ESSENTIAL LEED GA Exam Reference Guide

"This book is an invaluable tool in preparation for the LEED Green Associate (GA) Exam. As a practicing professional in the consulting realm, I found this book to be all-inclusive of the preparatory material needed for sitting the exam. The information provides clarity to the fundamental and advanced concepts of what LEED aims to achieve. A tremendous benefit is the connectivity of the concepts with real-world applications.

"The author, Gang Chen, provides a vast amount of knowledge in a very clear, concise, and logical media. For those that have not picked up a textbook in a while, it is very manageable to extract the needed information from this book. If you are taking the exam, do yourself a favor and purchase a copy of this great guide. Applicable fields: Civil Engineering, Architectural Design, MEP, and General Land Development."
—Edwin L. Tamang

Building Construction

**Project Management, Construction Administration, Drawings, Specs, Detailing Tips, Schedules, Checklists, and Secrets Others Don't Tell You
(Architectural Practice Simplified, 2ⁿᵈ edition)**

Learn the Tips, Become One of Those Who Know Building Construction and Architectural Practice, and Thrive in the Construction Industry!

For architectural practice, building design, and the construction industry, there are two kinds of people: those who know, and those who don't. The tips about building design, construction, and project management have been closely guarded by those who know—until now.

Most of the existing books on building construction and architectural practice are too expensive, too complicated, and too long to be practical or helpful. This book simplifies the process to make it easier to understand and uncovers the secrets of building design, as well as construction and project management. It sets up a solid foundation and fundamental framework for this field. It covers every aspect of architectural practice in plain and concise language that makes the information accessible to all people. Through practical case studies, the text demonstrates the efficient and proper ways to handle various issues and problems that arise in architectural practice, building design, and the construction industry.

The book is for ordinary people, aspiring young architects, as well as seasoned professionals in the construction industry. For ordinary people, it uncovers the tips of building construction; for aspiring architects, it works as a construction industry survival guide and a guidebook to shorten the process of mastering architectural practice and climbing up the professional ladder. For seasoned architects, it has many checklists to refresh the memory. It is an indispensable reference book for ordinary people, architectural students, interns, drafters, designers, seasoned architects, construction administrators, superintendents, construction managers, contractors, and developers.

You will learn:
1. How to develop your business and work with your client.
2. The entire process of building design and construction, including programming, entitlement, schematic design, design development, construction documents, bidding, and construction administration.
3. How to coordinate with government agencies, such as a county's health department or a city's planning, building, fire, or public works department.
4. How to coordinate with your consultants, including soils, civil, structural, electrical, mechanical, plumbing engineers, landscape architects, etc.
5. How to create and use your own checklists to provide quality control of your construction documents.
6. How to use various logs (i.e., RFI log, submittal log, field visit log, etc.) and lists (contact list, document control list, distribution list, etc.) to organize and simplify your work.
7. How to respond to RFI issues, CCDs, and review change orders, submittals, etc.
8. How to make your architectural practice a profitable and successful business.

Planting Design Illustrated

One of the most significant books on landscaping!

This is one of the most comprehensive books on planting design. It fills in the blanks in this field and introduces poetry, painting, and symbolism into planting design. It covers in detail the two major systems in planting design: formal planting design and naturalistic planting design. It has numerous line drawings and photos to illustrate the planting design concepts and principles. Through in-depth discussions of historical precedents and practical case studies, it uncovers the fundamental design principles and concepts, as well as underpinning philosophy for planting design. It is an indispensable reference book for landscape architecture students, designers, architects, urban planners, and ordinary garden lovers.

What Others Are Saying About *Planting Design Illustrated* ...

"I found this book to be absolutely fascinating. You will need to concentrate while reading it, but the effort will be well worth your time."
—Bobbie Schwartz, former president of APLD (Association of Professional Landscape Designers) and author of *The Design Puzzle: Putting the Pieces Together*.

"This is a book that you have to read, and it is more than well worth your time. Gang Chen takes you well beyond what you will learn in other books about basic principles like color, texture, and mass."
—Jane Berger, editor & publisher of gardendesignonline

"As a long-time consumer of gardening books, I am impressed with Gang Chen's inclusion of new information on planting design theory for Chinese and Japanese gardens. Many gardening books discuss the beauty of Japanese gardens, and a few discuss the unique charms of Chinese gardens, but this one explains how Japanese and Chinese history, as well as geography and artistic traditions, bear on the development of each country's style. The material on traditional Western garden planting is thorough and inspiring, too. *Planting Design Illustrated* definitely rewards repeated reading and study. Any garden designer will read it with profit."
—Jan Whitner, editor of the *Washington Park Arboretum Bulletin*

"Enhanced with an annotated bibliography and informative appendices, *Planting Design Illustrated* offers an especially "reader friendly" and practical guide that makes it a very strongly recommended addition to personal, professional, academic, and community library gardening & landscaping reference collection and supplemental reading list."
—Midwest Book Review

"Where to start? *Planting Design Illustrated* is, above all, fascinating and refreshing! Not something the lay reader encounters every day, the book presents an unlikely topic in an easily digestible, easy-to-follow way. It is superbly organized with a comprehensive table of contents, bibliography, and appendices. The writing, though expertly informative, maintains its accessibility throughout and is a joy to read. The detailed and beautiful illustrations expanding on the concepts presented were my favorite portion. One of the finest books I've encountered in this contest in the past 5 years."
—Writer's Digest 16th Annual International Self-Published Book Awards Judge's Commentary

"The work in my view has incredible application to planting design generally and a system approach to what is a very difficult subject to teach, at least in my experience. Also featured is a very beautiful philosophy of garden design principles bordering poetry. It's my strong conviction that this work needs to see the light of day by being published for the use of professionals, students & garden enthusiasts."
—Donald C Brinkerhoff, FASLA, chairman and CEO of Lifescapes International, Inc.

Index

www.ingramcontent.com/pod-product-compliance
Lightning Source LLC
Chambersburg PA
CBHW041431050326
40690CB00002B/500